D0856909

THE
ASTRONOMERS
ROYAL

EMILY WINTERBURN

ROYAL OBSERVATORY GREENWICH

NATIONAL MARITIME MUSEUM

First published in 2003 by the
National Maritime Museum, Greenwich,
London, SE10 9NF.

ISBN 0 948065 47 8

2 3 4 5 6 7 8 9

A CIP catalogue record for this book is available
from the British Library.

Editorial, design and production by The Book Group.
Cover design by Mousemat Design Ltd.

Cover: Portrait of George Biddell Airy by John Collier, © NMM,
London, reproduced by kind permission.

Printed and bound in China.

www.nmm.ac.uk/publishing

Contents

Foreword   4

A New Observatory for England   5

John Flamsteed   7

Edmond Halley   11

James Bradley   15

Nathaniel Bliss   19

Nevil Maskelyne   21

John Pond   24

Sir George Biddell Airy   27

Sir William Christie   30

Sir Frank Dyson   34

Sir Harold Spencer Jones   37

Sir Richard van der Riet Woolley   40

Sir Martin Ryle   44

Sir Francis Graham-Smith   46

Sir Arnold Wolfendale   48

Sir Martin Rees   50

The Royal Observatory, Cape of Good Hope   52

The Royal Observatory, Edinburgh   55

The Directors   58

Three Hundred Years of Astronomy   61

Further Reading   64

Picture Acknowledgments   64

Foreword

The Royal Observatory gives Greenwich a unique status on every map: all longitudes are calculated with reference to the Prime Meridian, which divides East from West. This book offers a succinct panorama of the individuals who moulded the Observatory's evolution over more than three centuries.

Astronomy is a venerable science – more so than any other apart from medicine – and perhaps the first to do more good than harm. Even in the seventeenth century, astronomers required large equipment and their work was recognized to be of strategic importance, especially for navigation. Mindful of this need, Charles II established the Royal Observatory in 1675. It was not until much later that other sciences were similarly supported by the government. That is why we have the antique title of Astronomer Royal, while there is no Chemist Royal, nor a Master of the Queen's Physicks.

In the twenty-first century astronomy is a more vibrant and rapidly advancing subject than it ever was, with wide public appeal. Within the UK, expertise is spread around dozens of universities and new technology offers ever-greater scope to schools and to the large community of amateur enthusiasts. These developments have changed the role of the Astronomer Royal: indeed, the last four holders of this title – since 1972 an honorary and rather anachronistic one – have been professors based in universities. Nonetheless, the original site of the Royal Observatory at Greenwich remains full of memories, and its long history offers inspiration to the millions who visit.

SIR MARTIN REES
JULY 2003

A New Observatory for England

King Charles II founded the Royal Observatory, Greenwich in 1675. The seventeenth century was a turbulent time for Britain. Guy Fawkes had tried to blow up the Houses of Parliament; there had been Civil War, the Great Plague and the Great Fire of London. Amidst all this, the Royal Society had been founded in 1660, providing for the first time a formal way for gentlemen interested in science to get together and discuss ideas. In France, too, science was becoming more popular and more structured. There, with King Louis XIV's backing, the Académie des Sciences was founded in 1665 and out of that a national observatory was set up in 1671.

The seventeenth century was also a time when international trade and colonization were becoming increasingly competitive. Tea and coffee were imported to Europe for the first time, while both France and Britain were building colonial outposts in North America. Travel by sea was the only way, certainly for Britain, to take part in these ventures, but navigation was extremely dangerous. The main problem was that without sophisticated navigation it was very difficult to know where you were, and this caused many ships to be wrecked on unexpected rocks or lose their crews to scurvy while taking too long to find land.

Although there were plenty of maps, knowing your exact position depended on being able to pinpoint your latitude (position north or south of the equator) and longitude (position east or west of an arbitrary point). While at sea latitude could easily be found from the height of the Sun above the horizon, working out your longitude remained a mystery.

In England in 1674 a Frenchman, the Sieur de St Pierre, was brought to the royal court by Louise de Kéroualle, the young French mistress of

Charles II. St Pierre had an idea that the longitude problem could be solved by measuring the changing position of the Moon against the stars. Philosophers at court, among them the astronomer Sir Christopher Wren (also known as the architect responsible for rebuilding London after the Fire), told Charles that while this might work in principle, star and Moon positions were too poorly known to make it possible.

So the King decided to establish a national observatory, like the one in Paris, and provided £500 to help finance the project. Unlike France, Britain did not have a tradition of lavish spending on science or, as it was then known, natural philosophy. Several sites were considered and eventually that of the ruinous 'castle' – an old hunting lodge – in Greenwich Park was chosen, in part because of the money that could be saved using the same foundations, and Wren was asked to design the building. The Observatory was built using spare bricks from Tilbury Fort, just east of London along the Thames, and lead, wood and iron from the demolished Coldharbour Gatehouse at the Tower of London, while labour costs were covered by the sale, for recycling, of decaying gunpowder from Portsmouth and the Tower.

In fact, the involvement of Louise de Kéroualle was simply a catalyst. The Royal Society had already been working on plans to build an observatory in the old King James I's College in Chelsea. Sir Jonas Moore, Surveyor-General of the Ordnance at the Tower of London had agreed to fund it and had suggested his protegé, the young astronomer John Flamsteed, as observer. When the King became involved these agreements were transferred to the new Royal Observatory at Greenwich and Flamsteed was appointed by Royal Warrant to be the King's 'Astronomical Observator', a title that gradually evolved into that of Astronomer Royal.

JOHN FLAMSTEED

Astronomer Royal, 1675–1719

The perfect star atlas

John Flamsteed was born near Derby in 1646. His father was a successful maltster in the brewing industry and his success meant that he could afford to send his son to university. Flamsteed was a rather sickly adolescent and this prevented him entering his chosen college, Jesus College, Cambridge, until he was 24. The delay, however, allowed him plenty of time for private study.

A solar eclipse in 1662 and another in 1668, which Flamsteed observed from Derby while convalescing, gave him the opportunity to get his name known in philosophical circles, by corresponding with a number of Fellows of the Royal Society. In 1670 he made the journey to London to meet some of these great philosophers. Among them was Sir Jonas Moore, who, recognizing his ability, became his patron. Moore supported him both financially, by providing him with astronomical instruments to help his work, and morally by pushing his name and ideas forward in discussions at court and at the Royal Society.

In 1674, without actually completing his degree, Flamsteed graduated (thanks to a little intervention by Jonas Moore) and the next year he took holy orders ready for a career in the Church. Before the nineteenth century training for the Church was one of the major reasons for going to university. For a budding astronomer it had the advantage of being a profession that could provide both an income and plenty of spare time for observing. For families that could perhaps only afford to send their sons to university but not support them after they graduated, it was an ideal arrangement.

JOHANNES FLAMSTEEDIUS Derbiensis

John Flamsteed, first Astronomer Royal.

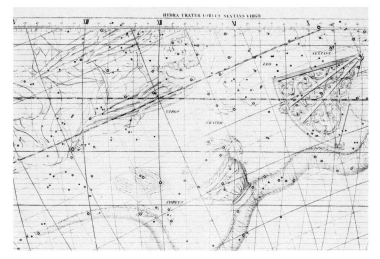

PLATE FROM FLAMSTEED'S STAR CATALOGUE.

In 1675, the 28-year-old Reverend John Flamsteed was appointed first Astronomer Royal. While the Observatory was being built at Greenwich, Flamsteed took up residence first in the Tower of London as a guest of Sir Jonas Moore, and then in the Queen's House, Greenwich, near the new Observatory site. The following year the Observatory was finished and Flamsteed moved in, setting up his telescopes in a shed at the end of the garden because the building itself, built by Wren and described in Wren's own words as being 'a little for pomp…', had neither an open roof nor windows properly aligned with a meridian. Both were essential for the job Flamsteed had been brought in to do.

His main task at the Observatory was to produce a map of the sky using a telescope – a project never before attempted. To do this he or his assistant would observe each star in the northern sky through a telescope as the star passed over the meridian (that is, a north–south line) on which the telescope was mounted. The scale on which the telescope was pivoted gave the angle of the star above the horizon. The other co-ordinate for the star was found by noting the exact time, by listening for the beats of a nearby pendulum clock, when the star passed over the meridian. Flamsteed spent the rest of his life making these observations. His days he filled taking in pupils (something he always resented) to supplement his income. All this time alone, working long hours with little company, made Flamsteed sometimes quite difficult to deal with.

Flamsteed was a very exact man. He prided himself on the accuracy of his results but was also very protective of them. The funding of the Observatory was such that most of the apparatus and any assistant's wages came out of either Flamsteed's pocket or that of his patron, Jonas Moore. For that reason, Flamsteed felt that the observations he made, and indeed the apparatus he used, belonged to him. So when Edmond Halley and Isaac Newton conspired together to get his unfinished work published in 1712, Flamsteed was furious. Indeed, he was so furious that when he managed to gain possession of 300 of the 400 copies printed, he went through each one, burning all but a few pages 'as a sacrifice to Heavenly Truth'.

Flamsteed's attitude towards his results and equipment had a number of repercussions for the Observatory. In the first place an accurate star catalogue, as requested by the King, was produced and published to Flamsteed's very high standards in 1725, six years after his death. In response to his reluctance to part with these results, however, a Board of Visitors was set up in 1710, with the authority to demand access to observations taken at the Observatory. Flamsteed's final legacy upon his death in 1719, was to have his wife strip the buildings of every piece of apparatus and furniture on the grounds that he had paid for it, leaving his successor to start again from scratch.

EDMOND HALLEY

Astronomer Royal, 1720–42

The man behind the Comet

Where Flamsteed had been a solitary and meticulous perfectionist, Edmond Halley was a charming and charismatic socialite. Halley was born in 1656 in a village just outside London, now part of the London borough of Hackney. His father was well-off, making a living as a landlord (although that income declined when many of his properties were destroyed in the Fire of London), and as a businessman making soap. His wealth allowed him to send Halley first to St Paul's School in London and then to Queen's College, Oxford.

At university Halley's family wealth allowed him to buy telescopes to develop his astronomical hobby, and he used the observations he made as a way of introducing himself to the newly appointed Astronomer Royal, John Flamsteed. Flamsteed was impressed by what he saw and encouraged Halley to write his first paper while still an undergraduate. Bored with university life, Halley came up with an idea to further his career and complement the work being carried out by Flamsteed. He wanted to travel to the southern hemisphere to catalogue the stars not visible from Greenwich. His father agreed to fund the trip, and Flamsteed used his position at court to secure him a place on an East India Company ship going to St Helena, an isolated British outpost off the coast of south-west Africa.

Halley dropped out of university in 1676 to sail for St Helena. He returned in 1678 and his *Catalogue of the Southern Stars* was published later that year. Rumours of Halley's sexual impropriety, both on this trip and on a later diplomatic visit to the astronomer Johannes Hevelius, had

EDMOND HALLEY, SECOND ASTRONOMER ROYAL,
BY SIR GODFREY KNELLER C. 1710.

a damaging effect on his relationship with the prim Flamsteed but for the time-being they remained on speaking terms. In 1682 Halley married Mary Tooke and together they moved into a house in Islington where he built himself a well-equipped private observatory, and he spent his time there and at the Royal Society. In that same year Halley began a correspondence with Isaac Newton over a newly visible comet (that would later take his name).

In 1684 tragedy struck. Halley's father was found murdered and he was dragged into a legal dispute with his stepmother over his father's estate. It was eventually settled amicably but nonetheless resulted in Halley having to give up his life of leisure and his fellowship of the Royal Society, and take up a paid position as the Society's clerk. He still managed to carry out a

THE GREAT AMATEUR.

AVIATOR. "MARVELLOUS FLIER! AND DOES IT FOR LOVE!"

OBSERVING HALLEY'S COMET, FROM A *PUNCH* CARTOON, 1910.

number of experiments, including inventing a diving bell in the 1690s to allow divers to breathe and carry out experiments under water. In 1691 he suggested a way in which the transit of Venus (when Venus is visible passing in front of the Sun's disc) could be used to calculate the size of the solar system and even predicted when the next two transits would occur. In 1698 he set sail on the ship *Paramore* to plot the variation in the Earth's magnetic field with a view to improving navigation.

Meanwhile, Halley's work on comets had made significant progress.

His work with Newton revealed that the latter had plenty of unpublished material that the world ought to see and Halley helped him put it all together in what would become Newton's most famous work, the *Principia*. Halley then started on his own investigations, looking back through historical records of eclipses and noticing certain patterns. From these, he concluded that comets travel round the Sun in an ellipse and are visible from Earth twice, once on their way towards the Sun and again as they travel away. It had previously been assumed that each viewing was of a separate comet. In 1705 he published his conclusions in *A Synopsis of the Astronomy of Comets* and added a prediction that there would be a return of the 1682 comet 'about the year 1758'.

In 1703 Halley became Savilian Professor of Geometry at Oxford, an appointment of which Flamsteed disapproved on the grounds that Halley 'now talks, swears, and drinks brandy like a sea captain', but which almost everyone else applauded. Then, on Flamsteed's death in 1719, Halley was appointed his successor at the age of 64. His main task as Astronomer Royal was to re-equip the Observatory, which he did with a grant he managed to obtain from the government, employing the best instrument makers in London to build him the two specialized telescopes needed for making maps of the sky: a transit instrument and an iron quadrant. Flamsteed had produced a map of the stars. The next task was to plot accurately the position of the Moon over an eighteen-year period called the Saros cycle – the time the Moon takes with its complicated motion to return to its original position in relation to the Earth and the Sun. While Halley survived to complete these observations his accuracy as an observer had clearly deteriorated with the onset of old age and the results were not of a high standard. His comet prediction was more successful. The comet of 1682 returned as he said it would in 1758, sixteen years after he died, and was named Halley's Comet.

JAMES BRADLEY

Astronomer Royal, 1742–62

Discoverer of the wobbly Earth

James Bradley was born in Sherbourne, Gloucestershire, in March 1693 to an aristocratic family. He was sent to Northleach Grammar School and then on to Balliol College, Oxford. His first contact with astronomy was through an uncle, the Reverend James Pound, an amateur astronomer and friend of both Halley and Newton, who helped to raise him. Like his uncle, Bradley went into the Church and became vicar at Bridstow, near Hereford in 1718, pursuing his interest in astronomy in his spare time.

In 1721 the post of Savilian Professor of Astronomy at Oxford became vacant and Bradley was appointed. A few years later, in 1725, while still professor at Oxford, Bradley began a long and fruitful collaboration with Samuel Molyneux, an astronomer of independent means. Molyneux came from a scientific Irish family, and married well, inheriting a house in Kew through his wife's family where he carried out most of his astronomical studies. While the Royal Observatory received government money to provide practical, utilitarian astronomy for the specific purpose of improving navigation, all other astronomy in England during the seventeenth, eighteenth and most of the nineteenth centuries was carried out by rich 'gentlemen amateurs' such as Molyneux.

At Molyneux's observatory in Kew, he and Bradley began to investigate 'parallax', the apparent displacement of a distant object when looked at from different angles. To investigate the parallax of a star, Molyneux and Bradley decided to observe it at two different times of the year. To do this they ordered a large zenith sector – a telescope that points straight

James Bradley, third Astronomer Royal.

up (to the zenith) and can move only a very small amount from side to side – from the instrument maker George Graham and began to observe the star Gamma Draconis. Their experiments failed to detect parallax but what Bradley discovered instead, when he installed a new zenith sector at his uncle's house in Wanstead, was the phenomenon of 'aberration'. Aberration is the slight displacement of a star caused by the fact that the Earth moves around its orbit at a certain speed while the speed of light is finite. It is an error that must always be kept in mind when making observations. This was his first great discovery.

Bradley succeeded Halley as Astronomer Royal in 1742, on Halley's recommendation. Like Halley, Bradley began his reign by re-equipping the Observatory. Halley's instruments remained and, thanks to his precedent, Bradley was able to get a grant to buy more, including the second quadrant that Halley had been denied. Bradley had an additional building constructed to house his new astronomical instruments and, like his predecessors, used the original Flamsteed House as living quarters.

He bought newer versions of the instruments used by Flamsteed and Halley, including a transit instrument which he set up on a new meridian – a meridian that continued to be used until 1851 and is still the basis for some some Ordnance Survey maps. He also installed new types of instruments, including his own zenith sector from home and some apparatus to investigate magnetism (as compasses were important for navigation) and meteorology (since changes in the weather affect the accuracy of telescopes).

Bradley's nephew, John Bradley, was appointed assistant and together they continued to carry out transit observations with the aim of improving and updating Flamsteed's results. As well as lunar observations, Bradley continued to work with his zenith sector. In 1748 he announced his second discovery using the instrument. This was the phenomenon of 'nutation', which is the slight wobble of the Earth on its axis caused by the shape of the Earth and its attraction to the Moon. This time the discovery won him the Royal Society's prestigious Copley medal.

While the primary concern of the Astronomers Royal was finding the exact position of stars to help navigators at sea, astronomers in the eighteenth century were generally interested in phenomena within the solar system, such as the passing of comets and eclipses. An important question at the time was the size of the solar system and this was thought to be solvable by observing the transit of Venus (as Halley had done). The planet Venus passes in front of the Sun's disc approximately twice every century and it was thought that if this could be observed simultaneously from more than one position on Earth then the observations could be used to find the distances between the Earth, Venus and the Sun and thus the size of the solar system. On 6 June 1761 Venus was due to transit the Sun for the first time that century and, although it was not something directly related to the official remit of the Observatory, Bradley was keen to be involved. In time, observing the transit of Venus would become a significant part of the Observatory's work. On this occasion Bradley was too ill to undertake the work and his friend Nathaniel Bliss took his place.

Bradley died in 1762 and there followed a long and confusing battle over the publication of his results. Like Flamsteed, Bradley had been a perfectionist and was unhappy to have his work released. After a long-running ownership dispute the results were finally printed in 1798, thirty-six years after Bradley's death.

NATHANIEL BLISS

Astronomer Royal, 1762–64

The shortest reign

Nathaniel Bliss was born in 1700 in Bisley, Gloucestershire, to a wealthy country gentleman and his wife. Little is known about his early life or where he went to school but he attended Pembroke College, Oxford, and

OIL PAINTING OF NATHANIEL BLISS, FOURTH ASTRONOMER ROYAL, BY AN UNKNOWN ARTIST C. 1755.

19

graduated in 1720. After completing his MA and taking holy orders, he became rector of St Ebbe's in Oxford until Halley died, whereupon he took up the vacant post of Savilian Professor of Geometry at Oxford. About this time Bliss struck up a friendship with Bradley, who succeeded Halley as Astonomer Royal. The two worked together on many projects over the years, sending each other their observations, and from this close-working relationship it was generally assumed that Bliss would be Bradley's successor at the Observatory.

Like Bradley, Bliss was also involved with the work of gentlemen amateurs. While Bradley had worked with Molyneux, Bliss contacted George Gerard, Earl of Macclesfield and owner of one of the best-equipped observatories in England, to work with him observing a newly visible comet. This appeared on 12 February 1745, and on 28 and 29 February the Earl took observations of the comet from his observatory at Shirburn Castle, while Bliss observed it from Bradley's base at Greenwich.

Bradley died in 1762 and, as predicted, Bliss was appointed his successor. However, Bliss did not outlive his friend by long and died only two years later, having made little impact on the observing programme of the Observatory. In fact things changed very little indeed as Charles Green, who had been Bradley's assistant and had carried out most of the actual observing at the Observatory, stayed on under Bliss in the same role.

Unlike under previous Astronomers Royal, ownership of the rights to the observations made at Greenwich under Bliss was relatively unproblematic. While the observations technically passed to his widow, she decided to sell and chose the Board of Longitude, the government body set up to award a prize to the person who solved the longitude problem, as the recipient.

Nevil Maskelyne

Astronomer Royal, 1765–1811

Harrison's enemy, Herschel's friend

Nevil Maskelyne was born in London in 1732. His family was wealthy and at the age of 9 he was sent to Westminster School. There he learned mainly classics but taught himself mathematics and astronomy from the new range of books then available, aimed at popularizing Isaac Newton's work. In 1749 he became a student at Trinity College, Cambridge, where he took his degree and holy orders. Staying on at Cambridge he became a fellow of his college in 1756 and his work there resulted in him being elected a Fellow of the Royal Society two years later.

Maskelyne's outstanding contributions to the Royal Society's journal, *Philosophical Transactions*, led the Society in 1761 to select him to sail to

Nevil Maskelyne, fifth Astronomer Royal.

St Helena (where Halley had gone almost a century before) to observe the transit of Venus, as Bliss was doing for Bradley in Greenwich. In the event, the weather in St Helena prevented Maskelyne from taking any observations, but his trip was not a total loss. On his journey to and from the island he carried out a number of experiments including testing some lunar tables by Tobias Mayer using a quadrant. This combination allowed him to measure his longitude at sea by the 'lunar distance method'.

In 1763, soon after he returned, Maskelyne published *The British Mariner's Guide*, explaining to sailors how to use the lunar distance method. A year later he was sent on another voyage, this time to Barbados, to test a rival method of finding longitude invented by a Mr Harrison. John Harrison was a clockmaker who for the previous 33 years had been working on a highly accurate and robust timepiece that could be used instead of star and Moon positions to find longitude at sea. The aim of this voyage was to test Harrison's fourth timekeeper, as the Board of Longitude required proof that it could keep time over a long sea passage. In the end the trial was considered inconclusive and Harrison continued to battle for his rights to the prize for another ten years. Maskelyne was appointed Astronomer Royal in 1765 soon after the four Harrison 'chronometers' were sent to the Observatory for safekeeping.

As Astronomer Royal, Maskelyne's first undertaking was to establish the *Nautical Almanac*, an annual publication giving the star, Moon and planet positions that were needed for the lunar distance method. The first volume came out in 1766 and it is still published to this day. To produce the almanac Maskelyne hired assistants, called 'computers', often straight out of school, who made the routine calculations on the observations to turn them into figures useful for mariners.

In his new role, Maskelyne also changed the buildings of the Royal Observatory, adding an extra room to what is now the Meridian

Building, extending the living quarters of the original Flamsteed House and changing the summer houses on either side of it into telescope domes for observing comets and planets.

In 1769 there was another transit of Venus and astronomers were keen to observe it again. This time Maskelyne wrote out instructions for observers and played an organizational role in arranging expeditions from England to locations around the world. Many of those sent had connections with the Observatory, including Charles Green, who joined Lieutenant James Cook on *Endeavour* to Tahiti. This time the results were generally successful, apart from a phenomenon called the 'black-drop effect' which made it difficult to determine the exact time at which Venus first passed across the Sun's disc.

While committed to continuing the Observatory's work for navigation, Maskelyne was also keen to keep up with modern developments in astronomy. In the 1770s he managed to persuade the Royal Society to fund an expedition to the mountain called Schiehallion in Perthshire, Scotland, to carry out an experiment known as 'weighing the Earth' to test Newton's law of universal gravitation. A few years later he supported a then unknown astronomer, William Herschel, using his contacts in England and on the continent to promote Herschel's discovery of the planet Uranus in specialist circles.

Maskelyne died in 1811 and, thanks to the *Nautical Almanac*, his work as Astronomer Royal was freely available to all.

H3, ONE OF FOUR TIMEPIECES JOHN HARRISON DESIGNED TO SOLVE THE LONGITUDE PROBLEM.

John Pond

Astronomer Royal, 1811–35

An early starter

Little is known of John Pond's family except that they must have been at least moderately wealthy, given his activities and his procrastination in finding paid work. Pond studied at Maidstone Grammar School and then took lessons from William Wales, a one-time computer of Maskelyne's, and one of the astronomers sent by him to observe the transit of Venus in 1769. At the age of 15 Pond was sufficiently confident of his own observations to suggest that those made at the Royal Observatory were inaccurate.

TRANSIT ROOM RE-EQUIPPED WITH JOHN POND'S INSTRUMENTS.

In 1783 Pond entered Trinity College, Cambridge, but left due to ill health before completing his degree. He then spent several years travelling around the Mediterranean and the Middle East. This was not altogether unusual; in the eighteenth century travel was considered to be a central part of the young English gentleman's education. The Grand Tour, as it was often termed, was to 'enrich the mind', 'rectify the judgement' and 'remove the prejudices of education'. Previously, Astronomers Royal had followed their university education with training for the Church. Pond, however, like Halley, appears to have had no immediate need for employment. He instead found himself being trained to be a well-rounded, well-educated English gentleman.

On his return to England in 1798 Pond set up his own private observatory in Westbury, near Bristol, using telescopes made by William Herschel and by the instrument maker Edward Troughton, who was by this time making a name for himself equipping observatories around the world. At this observatory Pond continued with the type of astronomy he had learned under William Wales, making observations comparable to those made at Greenwich. These observations confirmed that the instruments at Greenwich, which had remained unchanged for many years, were becoming deformed with age and as a result losing accuracy. This observation, coupled with his work with Troughton, who was also a friend of Maskelyne's, brought him to the attention of the Astronomer Royal and led to his nomination as his successor.

When Maskelyne died in 1811 Pond succeeded him as Maskelyne had wished. In his new role Pond immediately set about replacing the old apparatus with more modern equivalents, many of which he commissioned from his old collaborator, Troughton. As well as replacing old with new, Pond added to the range of apparatus used at the Observatory and expanded its remit. Now, in addition to making accurate observations of star, planet and Moon positions, the Observatory provided for mariners on the nearby River Thames to test the accuracy of their on-board timepieces by letting them know when it was 1 p.m. exactly by a visual signal – a time-ball dropped down a mast on the roof of the Observatory. This was installed in 1833. Under Pond the Observatory also began making regular magnetic and meteorological observations and a new building was established for that work.

Pond had close ties with France despite the Revolution and subsequent war with England, and in 1817 was awarded the Lalande Prize of the French Academy of Science. (Maskelyne had similar links with France; he was awarded the medal of the *Institut National des Sciences et des Arts* in 1803.) Pond retired in 1835 due to ill health and died a year later at his home in Blackheath, near Greenwich.

The time-ball installed on top of Flamsteed House in 1919, replacing Pond's time-ball of 1833.

Sir George Biddell Airy

Astronomer Royal, 1835–81

The Victorian modernizer

George Biddell Airy was born in Northumberland in 1801. His father, William Airy, worked for HM Excise, his mother, Anne Biddell, was the daughter of a farmer from Playford near Ipswich. Airy went to the grammar school in Colchester but additionally received an informal education from his uncle, Arthur Biddell, a self-educated land agent with many interesting friends, including a number who worked in engineering. Arthur Biddell also paid for Airy to have private tuition alongside his own son and this was sufficient to get Airy into Trinity College, Cambridge, on a scholarship.

At Trinity Airy did extremely well and graduated as First Wrangler (top of his class). This secured him employment as a mathematics tutor, which in 1826 led to his appointment as Lucasian Professor of Mathematics at Cambridge. A few years later the better paid post of Plumian Professor of Astronomy and Director of the Cambridge Observatory came up and he took it. At the Cambridge Observatory he used a firm of engineers that his

VANITY FAIR's 1875 CARICATURE OF GEORGE BIDDELL AIRY, SEVENTH ASTRONOMER ROYAL.

ASTRONOMERS PREPARING FOR THE 1874 TRANSIT OF VENUS EXPEDITION.

uncle knew, Ransomes of Ipswich, to build him a mounting to his own design for his new 'Northumberland' telescope, named after the Duke who provided the funds. He also began to produce annual reports of the work carried out by the Observatory to explain how money was spent each year.

In 1835 John Pond died and Airy was chosen to succeed him as Astronomer Royal. Unlike his predecessors Airy had no private income, or even the supplementary earnings of a curacy, and so he declined the position until the salary had been considerably increased. As Astronomer Royal he brought great organizational skills to the Observatory. He increased the number of computers (clerks who made the astronomical calculations) and established a clocking-in system to control their hours of work. He ordered new instruments, often to his own design, including a new transit circle, built by the Ipswich firm – by this time operating under the name of Ransomes and May – along with the instrument makers, Troughton and Simms. With this transit circle, Airy established a new meridian for Greenwich, one that would be adopted as the Prime Meridian (0° longitude) for the world in 1884.

As well as increasing staff numbers, Airy expanded the Observatory by reinterpreting its original remit and introducing new departments and new technologies. The work now included observing the rising number of newly discovered asteroids, recording regular magnetic and meteorological observations and making regular observations of the Sun's surface to record sun spots, which relate to variations in the Earth's magnetic field.

In 1845 Airy notoriously missed an important discovery. John Couch Adams in England and Urbain Leverrier in France simultaneously calculated the existence of a planet beyond Uranus and asked nearby observatories to look for it. Adams approached Airy at a time when Airy was perhaps not able to give him his full attention. The Observatory was in the middle of a crisis: Richardson, one of his assistants was being investigated on charges of murder and incest. Airy told Adams that he had neither the resources nor a suitable telescope and that such a search was not in the Observatory's remit. He did suggest that Adams ask his supervisor at Cambridge, Professor James Challis, to look using the Northumberland telescope. Adams, however, did not ask, or Challis did not respond: either way, Leverrier beat him to the discovery.

In his 46 years as Astronomer Royal Airy did not confine himself or his staff entirely to the Observatory. He also organized a number of expeditions. They included an updated version of Maskelyne's 'weighing the Earth' experiment, this time at Harton Colliery in South Shields, Tyne and Wear; a more rigorous transit of Venus project in 1874, sending out five sets of astronomers to observe the transit, each with identical equipment and instructions; and a number of eclipse expeditions – one to Turin in 1842, one to Sweden in 1851 and one to Spain in 1860. He continued Pond's work on time distribution, introducing the service that sent out Greenwich time by telegraph around the country. In 1880, Greenwich Mean Time, as it became know, was given Royal Assent as British Standard Time.

Sir William Christie

Astronomer Royal, 1881–1910

An international man of science

William Henry Mahoney Christie was born in Woolwich on 1 October 1845 to a wealthy and successful scientific family. His father was Professor of Mathematics at the Royal Military Academy in Woolwich and Secretary of the Royal Society 1837–54. His grandfather was the founder of the famous firm of auctioneers that still bears the family name.

William Christie, eighth Astronomer Royal.

Christie was educated at King's College, London and Trinity College, Cambridge, where he graduated as Fourth Wrangler (fourth in his class) in 1868. He was a fellow of Trinity for a year before taking the post of Chief Assistant to the Astronomer Royal, Airy, at Greenwich. In addition to his duties as Chief Assistant, Christie established a new magazine, *The Observatory*, which is still published.

In 1881 Airy retired and Christie was appointed in his place. As Astronomer Royal, the changes he implemented were vast and visually dramatic. The most striking of them was to the look of the Observatory. He added a total of eight new buildings, more than doubling the site area. Some buildings he was happy to justify to the Trustees; for others he was more artful. The New Physical Laboratory, now generally referred to as the South Building, was created very cautiously. Christie knew he would not get funding for such a large building and so designed it to be built in stages. The south wing, for chronometric work, was built first and was finished in 1894. The north wing followed in 1896 for magnetic and meteorological work. Then the east wing was built to accommodate the library and, finally, the west wing, for astro-photographic work, was completed in 1899.

Aside from the building work, Christie transformed the role played by the Observatory internationally by taking part in a new project. In 1887 a conference was organized at the Paris Observatory by Admiral Mouchez, David Gill of the Cape Observatory, Cape of Good Hope and Otto Struve, Director of the Pulkovo Observatory, Russia, to launch a collaborative project called the *Carte du Ciel* which aimed to produce a photographic map of the entire sky. The project brought together eighteen observatories from around the world. Each was allocated a section of the sky and required to produce a certain number of plates, to an agreed standard and magnification using standardized equipment, again agreed at this conference. This was the first time all these observatories had worked together as partners and their number included participants in Europe, Africa, South America and Australia.

The International Meridian Conference in Washington DC in 1884 was another example of the growing international flavour of the Observatory's work towards the latter end of the nineteenth century. Since Flamsteed's day, and even more so since Maskelyne had started to produce the *Nautical Almanac*, Greenwich had been used by a growing number of navigators as the reference meridian. Not everyone used the Greenwich meridian, however, and as international travel and trade increased this was becoming confusing. Similarly, working out the time difference between different parts of the world was problematic as there was no system in place to establish how these related to each other. The conference proposed to solve both these problems – by selecting a single or 'prime meridian' for the whole world and using this not only as 0° longitude but also as the start of the 'universal day' from which time around the world could be calculated. Votes were cast and the conference delegates agreed that Greenwich should be that site – home not only to the Prime Meridian but, as a consequence, also the baseline for world time, which Greenwich Mean Time thenceforward became.

Within the Observatory, too, Christie made great changes, expanding its scope to take in new developments in astronomy. In 1894 he installed a 28-inch telescope to look deep into space for double stars, or stars that can only be seen separately when observed at very high magnification. For a short time this was the largest refracting telescope in the world. It would have been way beyond the budgets of most amateurs and by this period most gentlemen amateurs were no longer able to keep up with new developments in research.

Christie continued to work on the various projects he had initiated at the Observatory until retiring in 1910. He died at sea on his way to Mogador, Morocco, on 22 January 1922.

The South Building being built about 1898.

Sir Frank Dyson

Astronomer Royal, 1910–33

Time and relativity

Frank Dyson was born in Leicestershire on 6 January 1868. He went to Bradford Grammar School and then on to Trinity College, Cambridge, where he graduated in 1889. After some time spent working at his old college, he was appointed Chief Assistant at the Greenwich Observatory in 1894. Since Airy introduced the post, the Chief Assistant, unlike the other assistant positions, was for someone with a university degree – ideally from Cambridge – who would act as deputy to the Astronomer Royal, managing the other assistants and taking charge of projects.

Frank Dyson, ninth Astronomer Royal.

Dyson's main task as Chief Assistant was to manage the *Carte du Ciel* project. In this he succeeded admirably and was able to take that experience and put it to good use in getting the same project underway in Edinburgh, when he became Astronomer Royal for Scotland in 1905 (see page 56).

Dyson returned to Greenwich in 1910 to take over from Christie, his former boss, as Astronomer Royal. His main interest was to improve the Greenwich Time Service, which by then offered time signals sent by telegraph to post offices across the country. He introduced new types of clocks to the Observatory and in 1924 pioneered the transmission of time signals through the newly invented medium, the wireless.

During the First World War the Observatory suffered a greatly reduced workforce as many members of staff were called up for military service. This was balanced to some extent by Belgian refugees. However, the reduction in staff and Admiralty demands to carry out work relevant to the war, including the testing of chronometers and binoculars, meant that much routine work was put on hold. As soon as the war was over, Dyson set about organizing a grand scientific venture. Well aware that there was a solar eclipse due to take place in 1919, he spent the last months of the war making arrangements with Arthur Eddington at Cambridge University for two expeditions, one to Sobral in Brazil, the other to Principe in West Africa to test Einstein's General Theory of Relativity. They would do this by detecting how much the light from a distant star was bent on its journey to Earth by the gravitational pull of the Sun. The expeditions were a success, providing the first-ever observational proof of Einstein's theory.

Back in Greenwich, Dyson began reinstating old routines and introducing new ones. One of the most widely known was the introduction of the 'six pips' – that is, the six beeps leading up to the hour on radio broadcasts of Greenwich Mean Time. Another change in the 1920s, this one more out of necessity than design, was to move the Magnetic Department to Abinger in Surrey, after electrification of the railway around Greenwich affected the magnetic measurements.

Harking back to the original work of the Observatory, and lurking in the basement ready to be uncovered by Rupert Gould, were the four Harrison chronometers. Gould had been visiting the Observatory in connection with some research he was doing on the history of the marine chronometer and had come across them quite accidentally. The Observatory had long since forgotten about them and they had been left to decay. Rather than trying to hide this unfortunate incident in its history, however, Dyson gave Gould full access to the Observatory archives and even allowed him to take the chronometers home, where he worked on restoring them to their former glory.

Dyson retired in 1933 and spent much of the inter-war period working with his equivalents at institutions around the world, trying to rebuild international relations in science. He died at sea on 25 May 1939 on his way to Australia.

ECLIPSED SUN, AS SEEN ON THE 1919 EXPEDITION TO SOBRAL.

SIR HAROLD SPENCER JONES

Astronomer Royal, 1933–55

Moving to Herstmonceux

Harold Spencer Jones was born in Kensington, London in 1890. He went to Latymer Upper School, a private school in nearby Hammersmith, and then on to Jesus College, Cambridge, where he graduated just before the First World War. From there he went immediately to start work as Chief Assistant at the Royal Observatory, Greenwich. During the war many of the staff were called up but Spencer Jones's expertise allowed him to carry out war work in England, as Assistant Director of Inspection of Optical Supplies for the Ministry of Munitions, testing equipment.

HAROLD SPENCER JONES, TENTH ASTRONOMER ROYAL.

In 1923 Spencer Jones was promoted from his job at Greenwich to HM Astronomer at the Cape of Good Hope (see page 54) following Sydney Samuel Hough's resignation. At the Cape Observatory he continued the work he had been doing at Greenwich on the *Carte du Ciel*, the international project both observatories were involved with. He also worked on another international study, this one initiated by the recently formed International Astronomical Union (IAU) to calculate the distance from the Sun to the Earth. This was essentially a new attempt to find the size of the solar system as the transit of Venus expeditions of the eighteenth and nineteenth centuries had done, this time with greater accuracy.

Dyson retired in 1933 and Spencer Jones returned to Greenwich. As Astronomer Royal he extended the work done by Dyson on time distribution, working with the Post Office to introduce the Speaking Clock in 1936. He also introduced quartz-crystal clocks, very accurate timepieces that would eventually show that the time determined by observing stars crossing the meridian (and thus determined by the rotation of the Earth) is irregular. That is to say, these clocks can keep time so well they showed that the time the Earth takes to make one rotation varies by a few seconds from year to year.

By 1938 London had expanded to engulf Greenwich. The Observatory was no longer in a village near the city but was in one of its suburbs. Street lighting, and pollution from factories and coal fires meant that the sky, already the frequent victim of poor weather conditions, was now almost impossible to observe. Spencer Jones decided that for the Observatory's work to be able to continue it would have to move to a new location. A site at Herstmonceux Castle in Sussex was found but the outbreak of war delayed the move.

All but a very few activities at the Observatory ceased during the

Second World War (1939–45). The big telescopes, including the 28-inch, were taken away and put into storage. This was just as well because the dome that had housed that telescope was severely damaged by bombing. Spencer Jones himself spent most of the war near the new Magnetic Department at Abinger in Surrey. Only the Solar and Meteorological Departments stayed on site and continued to observe.

SECOND WORLD WAR BOMB DAMAGE TO THE OBSERVATORY BUILDINGS.

After the war the move to Herstmonceux began, although progress was slow with the final pieces leaving Greenwich in 1957. While the relocation was underway, discussions took place as to the role of the old buildings and in 1951 they were handed over to the National Maritime Museum.

Before he retired in 1955, Spencer Jones chaired a committee set up by the Royal Society on the needs of British astronomy after the war. It was at this committee that plans were first formed to build a giant telescope that could be used by astronomers from all over Britain.

SIR RICHARD VAN DER RIET WOOLLEY

Astronomer Royal, 1956–71

A Commonwealth observer

Richard van der Riet Woolley was born in Weymouth, Dorset, on 24 April 1906, to a British father in the navy and a South African mother. He went to school in England and then moved to South Africa with his parents, where he took a degree at the University of Cape Town. He came back to England in the 1920s to study at Cambridge. There he took another degree and had the chance to work with Arthur Eddington investigating the atmosphere of stars, before gaining a fellowship to work in the USA at the Mount Wilson Observatory in California.

He returned to Cambridge in 1931 on a studentship and again worked with Eddington. On completion of his Ph.D he was appointed Chief Assistant at the Royal Observatory where, among other things, he collaborated with Dyson, then Astronomer Royal, on a book about eclipses. Ever ambitious, Woolley moved on quickly, returning to Cambridge when the opportunity arose to become Eddington's Chief Assistant at the University Observatory. Only a few years later, at the age of 33, he moved again, supported by his colleagues at Cambridge, to the post of Director of the Commonwealth Solar Observatory at Mount Stromlo in Canberra, Australia.

The observatory at Mount Stromlo was mainly concerned with observations and investigations into the nature of the Sun. When Woolley arrived the Second World War had just begun and much of the observatory, particularly the optics workshops, had been given over to helping the war effort by designing and developing gunsights. As Government Astronomer and in particular through his war work, Woolley got to know government

RICHARD VAN DER RIET WOOLLEY, ELEVENTH ASTRONOMER ROYAL.

ministers based in Canberra. After the war he used these well-nurtured relationships to help get his observatory equipped with the very best modern instruments. He was also keen to build up relationships with the new University in Canberra – the Australian National University – founded just before the war. Feeling that academic- rather than government-directed activity would improve the general level of research and international interest in the observatory, Woolley appealed to have the observatory transferred from government ownership to the University and eventually, after appealing directly to the Prime Minister, he succeeded.

In 1956 Spencer Jones retired from the Royal Observatory, Greenwich, or as it had then become known, the Royal Greenwich Observatory. When Woolley arrived to take up the position of Astronomer Royal, the Observatory was in the middle of moving from Greenwich to the clearer skies offered by Herstmonceux Castle in Sussex.

Unfortunately the disruption of the war and then the move had left the Observatory with old and out-of-date apparatus and working methods. In many ways it had not changed since Airy's day. Woolley set about rectifying the problem. The old emphasis of the Observatory on positional observations to aid navigators was outdated in the modern era. Christie had begun the break with the past and it was now down to Woolley to complete the process. While old routines did continue – there were still regular meridian observations to regulate time, for example – Woolley was keen to turn the Observatory into a research establishment to rival those now housed in the universities. He introduced astrophysics into its work, looking at the physics at work in stars and galaxies. He also worked on a project, initiated by his predecessor, of building a giant reflecting telescope called the Isaac Newton Telescope. It was decided that finding an outstation abroad for this would be prohibitively expensive so the telescope was built and housed at Herstmonceux.

HERSTMONCEUX CASTLE, WHICH BECAME THE NEW SITE OF THE OBSERVATORY.

While the Royal Greenwich Observatory (RGO) may not have been able to stretch to a new outstation of its own, Woolley had an alternative, inspired by his long-running connection with the Mount Stromlo Observatory. He suggested a joint collaboration with the Australians and in 1962 work began on building the large Anglo-Australian Telescope to explore the southern sky. Unfortunately, the costs involved made the continuing funding of the RGO's sister observatory at the Cape more and more difficult. To solve the problem, the Cape Observatory was handed over to the South African Government and, on his retirement from the RGO, Woolley became its first Director.

Sir Martin Ryle

Astronomer Royal, 1972–82

A radio astronomer at the Observatory

Martin Ryle was born in 1918 to an academic family. His father was a Professor of Social Medicine at Oxford, where Ryle took his first degree in physics, graduating in 1939. The networks created by scientists in the Second World War are well known. Many of the great names in modern physics first came together in the Manhattan Project working on the atomic bomb; many great mathematicians similarly formed close and long-lasting collaborations through their work at Bletchley Park on the Enigma code-breaking project. Ryle was lucky enough to join researchers at the Telecommunications Research Establishment at Malvern to work on radar at the outbreak of the war. It was this work at Malvern that would eventually earn him the Nobel Prize.

After the War Ryle was invited by one of his former Malvern colleagues to join him on a project at the Cavendish Laboratory in Cambridge, studying radio waves coming from the Sun and in effect producing a map of the Sun as seen in radio waves rather than visible light waves. This work helped him to secure a lectureship at Cambridge in 1948 followed by a fellowship at Trinity College and ultimately, in 1952, to become the first Professor of Radio Astronomy at Cambridge University. As part of his work at Cambridge, Ryle was involved with designing a telescope equivalent called a radio inferometer, for mapping the sky in radio waves.

In 1971, when Woolley retired as Astronomer Royal, the title became an honorary one with the running of the Observatory falling to the newly-created position of Director. As a result, when Ryle became

Astronomer Royal in 1972 he continued to be based at Cambridge. In 1974, while holding the title Astronomer Royal, Ryle received his Nobel Prize for physics for 'pioneering the new science of radio-astrophysics'.

MARTIN RYLE, TWELFTH ASTRONOMER ROYAL.

Sir Francis Graham-Smith

Astronomer Royal, 1982–90

The Manchester connection

Francis Graham-Smith was born in Roehampton, Surrey, in 1923. He went to Epsom College and then to Downing College, Cambridge, in 1941 to begin his university degree. Midway through the Second World War was a difficult time to start university and in 1943 his studies were interrupted when he was called upon to join the Telecommunications Research Establishment at Malvern to work on radar.

At Malvern Graham-Smith met Ryle, who had been there since the beginning of the war, and Graham-Smith joined his group to work on studying radio waves coming from the Sun. This placement gave Graham-Smith the opportunity to use scientific techniques as they developed in a brand-new field – an experience not open to most undergraduates. After the war he returned to Cambridge to finish his degree and was invited to join Ryle's group at the Cavendish Laboratory, using the radio inferometer to study radio waves from the Sun and later from other sources in the sky. In 1948 he collaborated on a

Francis Graham-Smith,
thirteenth Astronomer Royal.

paper with Ryle looking at the radio waves coming from a source in the constellation Cygnus.

Graham-Smith continued to investigate the exact source of the radio waves in this constellation and in the constellation Cassiopeía. In 1952 this led him to spend a year in the USA at the Carnegie Institute in Washington DC as a research fellow. Back in Cambridge he and Ryle began to look at how radio signals might be used in navigation and eventually came up with a theory suggesting that an accurate navigational system could be devised that used radio signals from an orbiting satellite. This idea eventually evolved into the Global Positioning System (GPS), now used to find both latitude and longitude at sea. It is an odd accident that because of the way the data was originally compiled, GPS shows the Prime Meridian a few metres east of Airy's meridian at Greenwich.

By the 1960s there were three main centres of radio astronomy research in Britain: Ryle's group at Cambridge, the group at Malvern, and Bernard Lovell's group at Jodrell Bank near Manchester. In 1964 Graham-Smith left Cambridge and went to Jodrell Bank, where he spent the next ten years studying radio waves from our galaxy and from the newly discovered pulsars – spinning, very dense stars called neutron stars, that emit a regular radio wave with each revolution – in a similar way to a lighthouse emitting a beam of light as its optic revolves. These were first discovered in 1967. He was also working at the Royal Greenwich Observatory, of which he became Director in 1976.

In 1981 he was offered the post of Director at Jodrell Bank and moved back to his first love of radio astronomy. A year later he was appointed Astronomer Royal but as the title no longer required an active role in the running of the RGO he was able to stay on in Manchester. Today he is Emeritus Professor at Manchester University and is still actively involved in the research carried out by the Pulsar Group at Jodrell Bank.

Sir Arnold Wolfendale

Astronomer Royal, 1991–95

Making astronomy popular

Arnold Wolfendale was born in Flixton, Greater Manchester, in 1927. He graduated from Manchester University with a degree in physics in 1948, making him the first Astronomer Royal to have been educated outside Oxford or Cambridge. He stayed in Manchester lecturing in particle physics and finishing his Ph.D. In 1956 he moved to the physics department of Durham University where he continued to work in the same field. This in turn led to work on the sub-atomic particles in space known as cosmic rays and more broadly on to cosmology – that is, the study of the Universe as a whole. In 1963 he became Professor of Physics at Durham University, before being made Head of Department in 1973.

Wolfendale has also held posts overseas. His first visiting lectureship, in 1952, was at the University of Ceylon, in what is now Sri Lanka. This was only a few years after independence, so a British academic might not have been the most obvious choice for the post. Astronomy, however, had by this time become an international activity. An exchange of academics between countries provides opportunities to share ideas and experience and, for new universities in new countries, offering visiting lectureships and professorships helped them to keep up to date with developments. In 1977 Wolfendale took another visiting post in Hong Kong and, after he had completed his term as Astronomer Royal, took another at the University of Yunnan in south-west China.

While still at the Physics Department at Durham, Wolfendale took on additional roles which would go on to shape the way in which he tackled his appointment as Astronomer Royal. Throughout the 1980s and 1990s he sat on various boards and committees involved in the funding and support of astronomy, and in all these posts he was active in promoting science to a wider audience than purely professional astronomers.

In 1991 he was given the honorary title of Astronomer Royal. Since its separation from the role of Director of the Royal Observatory, the post had become somewhat indefinable. Wolfendale gave it a purpose: he used his position and title to campaign actively and promote astronomy, both to the public to encourage their interest and enthusiasm, and to funding bodies to get more public money directed into research.

Wolfendale remained Astronomer Royal until 1995. In 1996 he was appointed Professor of Experimental Physics at the Royal Institution of Great Britain and since 1999 has been President of the European Planetary Society.

ARNOLD WOLFENDALE, FOURTEENTH ASTRONOMER ROYAL, BY HALLEY'S QUADRANT AT THE OBSERVATORY IN GREENWICH.

SIR MARTIN REES

Astronomer Royal, 1995–

Astronomer Royal for the twenty-first century

Born in England in 1942, Martin Rees went to Shrewsbury School and then on to Trinity College, Cambridge, where he graduated in mathematics in the 1960s. He became a fellow of Jesus College, Cambridge, but took time out in 1967 to work at the California Institute of Technology and in 1969 at the Institute of Advanced Studies at Princeton. It was during this period that his interests moved from mathematics to a developing area of astronomy – cosmology. At this time new discoveries, such as the so-called 'microwave background radiation', were providing the first observational evidence of the 'Big Bang'. This created new interest in questions surrounding the birth of the Universe; what the Universe is made of and how it has evolved over time. Rees began to work on these questions.

In 1972 Rees became a professor at Sussex University. But he soon returned to Cambridge, where, at the age of 31, he became the Plumian Professor of Astronomy and Experimental Philosophy, a post once held by Airy. In 1977 he added to this the title of Director of the Institute of Astronomy, also at Cambridge, and he held this post on and off until 1991. Like Wolfendale, Rees has taken many visiting posts in universities around the world, including Harvard University, the University of California and the University of Leiden. The work for which he is perhaps best known is his research (and the explanation of this work to non-scientific audiences) into the nature of such elusive cosmological concepts and objects as quasars, black holes, pulsars, star and galaxy formation, models of the Big Bang and dark matter.

Like Wolfendale, Rees has given considerable time and energy to promoting science to the general public and as Astronomer Royal since 1995, he has continued actively to pursue this work. He now works at the Institute of Astronomy, Cambridge.

Martin Rees, fifteenth Astronomer Royal.

The Royal Observatory,
Cape of Good Hope

By 1820, while the Royal Observatory at Greenwich had mapped the stars of the northern hemisphere for a century and a half with greater and greater precision, the south remained a comparative mystery. Halley had produced an early map of the southern sky from St Helena in 1678, while a little over half a century later, in 1752, the French astronomer, Abbé Nicholas Louis de Lacaille, had made observations from the Cape of Good Hope. However, there was no organization in place to produce regular observations of the southern-hemisphere sky.

In the years 1815–25 a small group of mostly Cambridge men, recognizing stagnation in many branches of science, established new societies and institutions to promote these disciplines. As undergraduates a number of them formed the Analytical Society at Cambridge with the aim of updating the Cambridge mathematics curriculum. The same group, with some additions, was also responsible for setting up the Royal Astronomical Society in 1820. In the 1810s, feeling that the interests of astronomy would be well served by having a permanent observatory in the southern hemisphere, this group set about making their case to the Royal Society and the Board of Longitude. They chose the Cape of Good Hope partly because it was then within the British Empire, partly because it was at approximately the same longitude as Greenwich, and partly because of the frequent fine weather.

Their proposal was successful, and in 1820 a new Royal Observatory was founded at the Cape of Good Hope, with Fearon Fallows, a Cambridge graduate and one of the founder members of the Analytical Society, as its first HM Astronomer at the Cape, a title sometimes shortened to 'the Astronomer'. To equip this new observatory, identical instruments to those used at Greenwich by Pond (who was at that time Astronomer Royal) were commissioned and shipped over.

For the first century of the Cape Observatory the Astronomer played a major part in the evolving programme and its links with its sister institution at Greenwich. At first all the Astronomers were picked and sent over from Britain. Fallows, who was responsible for finding the site and getting the observatory built and staffed, had graduated from Cambridge. Thomas Henderson, who took over when Fallows died of scarlet fever in 1831, was from Scotland. He stayed only a couple of years, however, describing the place as a 'Dismal Swamp' where he was expected to live 'among Slaves and Savages… [and] plenty of insidious venomous deadly snakes'. Thomas Maclear, who took over from him, was a friend of John Herschel, one of the key members of that original group of Cambridge men and son of Maskelyne's friend William Herschel. He took up the post at the Cape shortly before Airy was appointed Astronomer Royal at Greenwich. Maclear, like his Greenwich counterparts, Pond and Airy, extended the buildings of the Cape Observatory, installed a time-ball and continued the work of plotting the stars of the southern hemisphere.

Maclear retired in 1870 and was replaced by Edward James Stone. Stone had previously been Chief Assistant under Airy at Greenwich and came to the Cape on Airy's recommendation. While Maclear had been distracted with a major surveying project in the latter part of his time at the Cape, the buildings and the work of the observatory had been left to decline. Stone dedicated his nine years in office to turning the old observations into useful catalogues and doing the same with his new observations.

David Gill, a trained watchmaker and experienced amateur astronomer, replaced Stone in 1879. Before setting out he made a tour of foreign observatories to obtain ideas for the renovations he was required to carry out when he arrived at the Cape. Using many of the same instrument makers who supplied Greenwich, Gill set about introducing new technology, as Christie was doing back in England, such as spectroscopes for finding out what stars are made of, and photography. He stayed for 28 years and in that time the Cape Observatory was transformed from a place of utilitarian routine observations to one of innovation and research, on a par with any in the northern hemisphere.

The twentieth-century Astronomers made less dramatic changes individually to the running of the observatory, as astronomy became more and more an international affair with many projects run as collaborations with foreign partners, chief among them, of course, being Greenwich. Sydney Samuel Hough, Gill's successor, had been Gill's assistant at the Cape straight from university. He retired in 1923 and was replaced by Harold Spencer Jones, previously Senior Chief Assistant at Greenwich. John Jackson, who had worked with Spencer Jones as Junior Chief Assistant at Greenwich, took his place at the Cape when Spencer Jones returned to England as Astronomer Royal.

Finally, Jackson's chief assistant, Richard Hugh Stoy, replaced him as the Astronomer in 1950. In 1971 the Royal Observatory, Cape of Good Hope, closed. The very next day it reopened as the South African Astronomical Observatory with ex-Astronomer Royal from Greenwich, Richard Woolley, as its first director.

The Royal Observatory, Edinburgh

The Royal Observatory began as a local resource, set up by professors at Edinburgh University for the use of both students and local amateurs. On a visit to the city in 1822, King George IV was asked if he would give his patronage to the observatory. In 1834 the post of Astronomer Royal for Scotland was created with a Royal Warrant requiring him:

> To take upon himself the care and custody of all instruments within the Observatory of Edinburgh which belong to Us and to apply himself with diligence and zeal to making astronomical observations at the said Observatory for the extension and improvement of Astronomy, Geography and Navigation and other branches of Science connected therewith.

This was a much broader remit than Greenwich had been given on its foundation some 159 years earlier.

First in the post was Thomas Henderson, recently returned from his time at the Cape Observatory. Henderson was there for twelve years setting up and equipping the Edinburgh Observatory. He died in 1846 and was replaced by the enigmatic Charles Piazzi Smyth, astronomer, artist and expert on the metrology of the pyramids of Egypt. Piazzi Smyth, like Henderson, had worked at the Cape. He had been assistant to Maclear when only 16 years old and had become good friends with the Herschel family. On his return from the Cape, with John Herschel's help, he was appointed as Henderson's successor. Piazzi Smyth turned Henderson's observations into useful tables, gave lecture courses at the University, carried out his own observations of star positions and introduced a time service similar to that at Greenwich. He also came up with the idea of building an observatory on a mountain in Tenerife and even obtained money to go on an expedition to explore the viability of such a project. Although the trip went well, the idea was put on hold and forgotten about until the late-twentieth century. Various other projects distracted him in his later years, including his fascination with the pyramids, and in 1888 he retired from office leaving the observatory in a poor state.

The observatory was in such a state of disrepair in 1888 that serious consideration was given to abolishing the post of Astronomer Royal for Scotland and turning the buildings into part of the University. James Lindsay, 26th Earl of Crawford, came forward to protect the institution. He provided money and equipment to build a new observatory to replace the original building at Calton Hill. Ralph Copeland was appointed Astronomer Royal and given the task of finding a new site and returning the institution to its former glory.

When Copeland left in 1905 the observatory was back on its feet. It had new buildings at Blackford Hill and a revised observing programme that included spectroscopy (a technique for investigating the chemical composition of stars) and photography, as well as the standard star-position observations. Frank Dyson, a former assistant at Greenwich, was Copeland's successor but stayed at the observatory only five years before moving back to Greenwich to take up the post of Astronomer Royal.

As at the Cape, the Edinburgh Observatory in the twentieth century was mainly involved in international collaborations in astronomy. Ralph Allan Sampson, who replaced Dyson, updated the apparatus, introducing more that could be used to investigate the chemical composition of stars. William Greaves, who took over in the inter-war years, had, like Dyson, started out as an assistant at Greenwich. His particular interest was to investigate the temperature of stars and he brought this to the work of the

PIAZZI SMYTH'S EXPEDITION TO TENERIFE, 1856.

observatory. Hermann A. Brück replaced Greaves in 1957. He introduced new electronic technology to every aspect of the observatory's work and began a programme of studying galaxies, using observations taken there years earlier. Brück was followed in 1975 by Vincent Reddish, who continued many of his projects, and then by Malcolm Longair in 1980, who did the same.

The current Astronomer Royal, John C. Brown, was the first to take on the role after it was separated from the post of Regius Professor of Astronomy at Edinburgh University. He distinguishes between the research work he does in astronomy for the University and his role as Astronomer Royal in improving the public appreciation of astronomy, which includes planetarium shows and talks to schools and the general public. When the Royal Greenwich Observatory closed in 1998 Edinburgh took over some of its astronomical work.

The Directors

Ask anyone connected with the old 'working' Royal Greenwich Observatory (RGO) why the role of Astronomer Royal was split in two after Richard Woolley retired in 1971 and you will get one in a range of highly conspiratorial answers. Some will tell you it was to avoid giving the title Astronomer Royal to the most appropriate candidate, who was a woman; others that it was to appease the universities, who were jealous of the resources given to the Observatory. Whatever the true reason, in 1971 the title of Astronomer Royal became purely honorary, while the actual day-to-day running of the Observatory and development of its research programme was carried out by a Director.

The first Director was Margaret Burbidge. She had studied at the

THE ISAAC NEWTON GROUP OF TELESCOPES AT LA PALMA.

University of London and worked at observatories in Chicago, Cambridge and California before being appointed Director in 1972. She stayed only a year at the Observatory before moving back to the USA. At Herstmonceux, Burbidge was replaced by Alan Hunter. As Director, Hunter continued to oversee the routine work and research programmes established by Woolley, and to work on arrangements for the tercentenary celebrations in 1975. He was replaced in 1976 by future Astronomer Royal, Francis Graham-Smith, whose time as Director was divided between running the Observatory itself and work on the early stages of the Northern Hemisphere Observatory at La Palma in the Canary Islands. This had initially been proposed in the late 1960s by the Astronomer Royal for Scotland, Hermann Brück, following his predecessor, Charles Piazzi Smyth's, investigations a century earlier into the advantages of doing

astronomy from a point high above sea level in a good climate. It later became an international observatory, run jointly by astronomers from all over Europe. It houses, among others, the Isaac Newton Telescope that began life at Herstmonceux. The observatory is generally known as the Isaac Newton Group of telescopes, or ING.

Graham-Smith left in 1981 and was replaced by Alec Boksenberg. Under Boksenberg the work of the RGO became increasingly focused on the La Palma site and a good deal of time and expertise was given over to designing and constructing a second telescope there. Boksenberg was the last Director of the Observatory at Herstmonceux. In 1990 it was decided to move Observatory staff to Cambridge. Jasper Wall was the very last Director, appointed in 1995, a few years after the move, and was given the unpleasant task of winding up activities at Cambridge ready for its closure in 1998.

Three Hundred Years of Astronomy

By the time of its closure in 1998, the Royal Greenwich Observatory, a modern research institution, would have been utterly unrecognizable to its seventeenth-century founders. Although the fifteen men who have held the title Astronomer Royal over the last 300 years have some things in common, such as their frequent close association with Oxford and Cambridge Universities, it is because of their differing approaches to running this great institution that the Observatory was able to evolve and keep up with the times.

In the eighteenth century, the Observatory was mainly used to provide raw data for navigators while major discoveries were made by rich and leisured gentlemen-amateur astronomers. Nonetheless, the Astronomers Royal played a part in developing the Observatory, setting the standards it could achieve within its narrow remit. John Flamsteed, as the first postholder, defined the role, producing a catalogue of stars of the highest accuracy. Edmond Halley used his charm to get the first grant for equipping the Observatory with new apparatus, setting a precedent for his successors. James Bradley used his position to extend the site with new buildings, while Nathaniel Bliss, during his short tenure but long association with the Observatory, was the first to link it to the great transit-of-Venus expeditions of the eighteenth and nineteenth centuries.

Thanks largely to its far-seeing Astronomers Royal, the nineteenth-century Observatory went through perhaps its most dramatic transformation, from gathering navigational information to pure research. Under Nevil Maskelyne the data-collection aspect of its work was formalized and published in the *Nautical Almanac*, leaving more scope to get involved with other research projects such as the transit of

Venus and the 'weighing the Earth' experiment. John Pond brought Flamsteed's old concerns about accuracy back to centre stage and, as a result, kept the apparatus up to date. George Biddell Airy expanded both staff numbers and the range of activities that could be linked to the Observatory's original remit and implemented structures that were to remain until the mid-twentieth century. William Christie took Airy's expansion one stage further, for the first time making a break with the original remit and bringing in purely astronomical research that had little or no connection with navigation.

In the twentieth century astronomy changed again. From looking at the position of objects in the sky in the eighteenth century, to the chemical composition of objects in the nineteenth, the new central theme in astronomy in the twentieth century became the application of modern physics. As before the Observatory kept up to date through the work of its Astronomers Royal. Frank Dyson brought the Observatory back in line with its original remit providing Greenwich Mean Time to the nation. His successor, Harold Spencer Jones, moved the Observatory to a place where the skies were clearer, while Richard Woolley brought the Observatory right into the modern era, breaking entirely from the original remit and introducing the new field of astrophysics. After Woolley, the title Astronomer Royal came to mean something different and it was down to the Directors to decide the direction of the Observatory's research. Observation itself was then moved out of Britain to higher, clearer locations such La Palma and Australia under the Observatory directors in the 1970s, 1980s and 1990s. The Astronomers Royal – Martin Ryle, Francis Graham-Smith, Arnold Wolfendale and Martin Rees – meanwhile have introduced new audiences to the science that has kept astronomers occupied at the Royal Observatory for over three hundred years.

Astronomers of the Royal Observatories

ROG = Royal Observatory, Greenwich
ROC = Royal Observatory, Cape of Good Hope
ROE = Royal Observatory, Edinburgh

Date	ROG	ROG director	ROC	ROE
1675	Flamsteed			
1720	Halley			
1742	Bradley			
1762	Bliss			
1765	Maskelyne			
1811	Pond			
1820			Fallows	
1831			Henderson	
1833			Maclear	
1834				Henderson
1835	Airy			
1846				Piazzi Smyth
1870			Stone	
1879			Gill	
1881	Christie			
1889				Copeland
1905				Dyson
1907			Hough	
1910	Dyson			Sampson
1923			Spencer Jones	
1933	Spencer Jones		Jackson	
1938				Greaves
1950			Stoy	
1956	Woolley			
1957				Brück
1972	Ryle	Burbidge		
1973		Hunter		
1975				Reddish
1976		Graham-Smith		
1980				Longair
1982	Graham-Smith	Boksenberg		
1991	Wolfendale			
1995	Rees	Wall		Brown

Further Reading (by chapter)

John Birks, *John Flamsteed, the first Astronomer Royal at Greenwich*, 1999.
Eric Forbes, Lesley Murdin and Frances Willmoth (eds), *The Correspondence of John Flamsteed* (3 vols), 1995, 1997, 2003.
Colin A. Ronan, *Edmond Halley, Genius in Eclipse*, 1969.
Alan Cook, *Edmond Halley, Charting the Heavens and the Seas*, 1998.
Derek Howse, *Nevil Maskelyne, the Seaman's Astronomer*, 1989.
Wilfrid Airy (ed.), *Autobiography of Sir George Airy*, 1896.
Margaret Wilson, *Ninth Astronomer Royal, the life of Frank Watson Dyson*, 1951.
Brian Warner, *Astronomers at the Royal Observatory, Cape of Good Hope*, 1979.
Brian Warner, *Royal Observatory, Cape of Good Hope, 1820–1831*, 1995.
Hermann A. Brück, *The Story of Astronomy in Edinburgh*, 1983.

Picture Acknowledgments